AI美男人物绘画关键词图鉴

Midjourney版

AIGC-RY研究所 著

人 民 邮 电 出 版 社
北 京

图书在版编目（CIP）数据

AI美男人物绘画关键词图鉴 : Midjourney版 / AIGC-RY研究所著. -- 北京 : 人民邮电出版社, 2024.2
ISBN 978-7-115-63194-7

Ⅰ. ①A… Ⅱ. ①A… Ⅲ. ①图像处理软件 Ⅳ. ①TP391.413

中国国家版本馆CIP数据核字(2024)第000039号

内 容 提 要

AI是当下无法阻挡的艺术创作趋势。

本书首先简要地介绍了一下关键词的使用方法，帮助读者大致了解生成图片的基本原理；正文实战部分，展示了“霸道总裁”、俊美男、优雅绅士、古风美男、美少年、魅力大叔、艺术家等7类美男主题的AI图片生成效果，并给出了提示词说明，通过图文对应的方式帮助读者了解生成图片的具体方法，从而生成自己想要的图像。

本书适合对AI图像创作感兴趣的读者和有AI图像创作需求的设计师、插画师等阅读。

◆ 著　　　　AIGC-RY 研究所
责任编辑　王　铁
责任印制　周昇亮
◆ 人民邮电出版社出版发行　北京市丰台区成寿寺路 11 号
邮编 100164　电子邮件 315@ptpress.com.cn
网址 https://www.ptpress.com.cn
北京九天鸿程印刷有限责任公司印刷
◆ 开本：700×1000 1/16
印张：9　　2024 年 2 月第 1 版
字数：230 千字　　2024 年 2 月北京第 1 次印刷

定价：49.80 元

读者服务热线：(010)81055296　印装质量热线：(010)81055316
反盗版热线：(010)81055315
广告经营许可证：京东市监广登字 20170147 号

本书使用说明

① 总关键词

本章用到的所有关键词

② 关键词

达到图片效果必须要用到的关键词

第1章

“霸道总裁”

“霸道总裁”，他们在商界取得了巨大成功，积累了巨额财富，在社会上享有很高的地位。同时，他们的自我认知让人无法忽视，且时常保持冷静和克制，使人难以洞察他们的情绪。虽然他们可能给人一种坚不可摧的印象，但他们也会有脆弱的情感，快来深入了解他们吧！

关键词：authoritative leader, ceo, commanding CEO, dominant CEO, tall and powerful, cold, tough, profound, mature, wise, elegant, handsome, confident, attent ① fidently, domineering CEO, charming and refined gentleman, atter ① confidence, admiring, captivating, meticulousness, thoughtful, affinity, emotional, meticulously, humorous, peaceful, scientific, talent, sensitivity, athletic, joy, patience, natural, charm, leisurely, adventurous

authoritative leader +tall and powerful ②

提示词：male, authoritative leader, tailored ③ s suit, confident gaze, tall and powerful, commanding presence, sharp jawline, gold glasses, ani ③ -ar 1:1 --niji 5

commanding CEO +cold

④

⑤

提示词：male, commanding CEO, polished suit and tie, captivating smile, enigmatic charm, cold, gold glasses, anime style --ar 1:1 --niji 5

6

③ 提示词

达到图片效果所需的全部提示词

④ 图片效果

输入提示词会出现的相关图片效果

⑤ 重点提示词

提示词中的重点提示词汇（即关键词）用彩色标注出来，一目了然

目录 CONTENTS

第1章

“霸道总裁”

“霸道总裁”，他们在商界取得了巨大成功，积累了巨额财富，在社会上享有很高的地位。同时，他们的自我认知让人无法忽视，且时常保持冷静和克制，使人难以洞察他们的情绪。虽然他们可能给人一种坚不可摧的印象，但他们也会有脆弱的情感，快来深入了解他们吧！

关键词：authoritative leader, ceo, commanding CEO, dominant CEO, tall and powerful, cold, tough, profound, mature, wise, elegant, handsome, confident, attentive, confidently, domineering CEO, charming and refined gentleman, attentively, confidence, admiring, captivating, meticulousness, thoughtful, affinity, emotional, meticulously, humorous, peaceful, scientific, talent, sensitivity, athletic, joy, patience, natural, charm, leisurely, adventurous

authoritative leader +tall and powerful

提示词：male, authoritative leader, tailored business suit, confident gaze,tall and powerful,commanding presence, sharp jawline, gold glasses, anime style --ar 1:1 --niji 5

commanding CEO +cold

提示词：male, commanding CEO, polished suit and tie,captivating smile, enigmatic charm， cold, gold glasses, anime style --ar 1:1 --niji 5

authoritative leader+cold +confident

提示词：male, authoritative leader, tailored business suit, a cold expression, confident, anime style --ar 1:1 --niji 5

authoritative leader+tall and powerful+cold + confident

提示词：male, authoritative leader, tailored business suit, tall and powerful, a cold expression, confident, anime style --ar 1:1 --niji 5

commanding CEO+confident+wise+elegant+handsome

提示词：male, commanding CEO, polished suit and tie, confident,wise, elegant, handsome, gold glasses, anime style --ar 1:1 --niji 5

commanding CEO+elegant+confident+handsome

提示词：male, commanding CEO, polished suit and tie, enigmatic charm,elegant, confident, handsome, charming smile,anime style --ar 1:1 --niji 5

commanding CEO+mature+tall and powerful+cold

提示词 : male, commanding CEO,polished suit and tie,mature,a tall and powerful figure, confident, cold, anime style --ar 1:1 --niji 5

authoritative leader+cold+elegant

提示词 :male,authoritative leader,tailored business suit,strong facial features,cold,elegant, anime style --ar 1:1 --niji 5

CEO+tough+profound+mature

提示词：male, CEO confidently struts through a sleek and modern office building,tough,profound, mature, the floor to ceiling windows showcasing a breathtaking cityscape with towering skyscrapers,anime style --ar 1:1 --niji 5

CEO+elegant+handsome+confident

提示词：male, sitting posture,charismatic CEO, elegant, handsome, confident, the floor-to-ceiling windows showcasing a breathtaking cityscape with towering skyscrapers,anime style --ar 1:1 --niji 5

dominant CEO+cold+tough+profound+mature+wise

提示词：male, a dominant CEO, dressed in a sleek black suit, stands confidently in a modern glass-walled office, a futuristic cityscape at sunset,cold, tough, profound, mature, wise,anime style --ar 1:1 --niji 5

dominant CEO+profound+cold+mature+wise

提示词：male, a dominant CEO, reclines on a armchair in a private library filled with ancient books, profound, cold, mature, wise,anime style --ar 1:1 --niji 5

CEO+tall and powerful+profound+cold

提示词：male, CEO stood in a customized suit, there are scattered bright lights in the dim sky, creating a mysterious atmosphere, tall and powerful, profound, cold,anime style --ar 1:1 --niji 5

CEO+tall and powerful+profound

提示词：male, charismatic CEO, dressed in a sleek tailored suit, with sharp features and piercing eyes, tall and powerful, profound, stands against a backdrop of a opulent penthouse suite,anime style --ar 1:1 --niji 5

CEO+cold+tough+profound+mature

提示词：male, CEO, dressed in a tailored black suit, sits in a luxurious leather chair in a boardroom filled with sleek,high-tech gadgets,cold, tough, profound, mature,anime style --ar 1:1 --niji 5

dominant CEO+mature+wise+elegant+handsome

提示词：male, sedan rear seat,a dominant CEO, dressed in an impeccably tailored black suit,mature, wise, elegant, handsome,anime style --ar 1:1 --niji 5

CEO+mature+wise

提示词：male, CEO,with the dazzling city night view beyond the floor-to-ceiling windows, mature, wise, a man in sharp tailored attire, his gaze exudes unwavering confidence,anime style --ar 1:1 --niji 5

CEO+elegant+handsome+confident

提示词：male, CEO, amidst a lavish executive suite, illuminating a sleek, expansive desk,elegant,handsome, confident,anime style --ar 1:1 --niji 5

CEO+elegant+handsome+confident

提示词：male, in the vast garden of the private villa, the sun shines, a tall and majestic CEO sits on a beautifully carved chair, elegant,handsome,confident,anime style --ar 1:1 --niji 5

CEO+elegant+handsome+confident

提示词：male, CEO, elegant, handsome, confident,in the rooftop in the evening, stands a commanding corporate executive, clad in a sharp suit,anime style --ar 1:1 --niji 5

CEO+handsome+confident

提示词：male, CEO, leaning casually against a sleek luxury car, the setting sun casts a golden hue, mirroring the aura of authority that surrounds him,handsome, confident, anime style --ar 1:1 --niji 5

CEO+wise+elegant+handsome

提示词：male, CEO sits at the head of the table, radiating an aura of unwavering authority, wise, elegant, handsome, anime style --ar 1:1 --niji 5

CEO+mature+wise

提示词：male, amidst the sophisticated ambiance of a dimly lit upscale restaurant, CEO,his poised demeanor and controlled gestures mirror his dominance as he navigates conversations with strategic precision, mature, wise, anime style --ar 1:1 --niji 5

CEO +confident

提示词：male, amidst the neon lights of the city, CEO, his tailored suit accentuates a powerful aura as he walks with confident determination, anime style --ar 1:1 --niji 5

CEO+confident+attentive

提示词：male, in side a private luxury plane, a CEO sits in a comfortable seat, confident, attentive, the clouds outside seemed to complement his foresight in the business world,anime style --ar 1:1 --niji 5

CEO+handsome+confident

提示词：male, close shot, CEO sat on customized furniture, in the spacious private villa, handsome, confident, anime style --ar 1:1 --niji 5

CEO+confidently

提示词：male, CEO dressed in a carefully matched suit, confidently stood inside the airport, his demeanor was calm, demonstrating his leadership and decision-making abilities,anime style --ar 1:1 --niji 5

domineering CEO+handsome+confident

提示词：male, in a peaceful resort, a domineering CEO sits on the balcony of a private villa, the magnificent sea view, handsome, confident,anime style --ar 1:1 --niji 5

the domineering CEO+mature+wise

提示词：male, the domineering CEO sitting on the deck of a luxury yacht demonstrated his ruling style, contemplating the future of commerce,mature, wise, anime style --ar 1:1 --niji 5

CEO+elegant+handsome+confident

提示词：male, CEO stood on the dinner stage, smiling and communicating with the guests,confident, every smile contained business opportunities, elegant, handsome, anime style --ar 1:1 --niji 5

the domineering CEO+confidence+mature+wise

提示词：male, wearing a delicate suit, the domineering CEO stood on the stage, mature, wise, his gaze sharp as he scanned the audience, his speech was full of authority and confidence, demonstrating absolute control over the business world, anime style --ar 1:1 --niji 5

CEO+mature+wise+elegant+handsome+confident

提示词：close shot, male, standing in the center of the conference room, CEO, which is not easy to disobey,mature, wise, elegant, handsome, confident,anime style --ar 1:1 --niji 5

CEO+wise+confident

提示词：male,the CEO sat in the open-air coffee shop,chess, his gaze was sharp, every word conveyed business strategy and market insight, wise,confident,anime style --ar 1:1 --niji 5

CEO+wise+confident

提示词：male, CEO sat in the open-air coffee shop, chatting with his partners, his gaze was sharp, and every word conveyed business strategy and market insight, wise,confident,anime style --ar 1:1 --niji 5

CEO+wise+confident

提示词：male, holding a microphone, CEO delivered a speech at an outdoor charity event, his words were full of charitable mission and social responsibility,wise,confident,anime style --ar 1:1 --niji 5

CEO+wise+confident

提示词：male, CEO,standing at the technology exhibition site, his vision is full of the future, demonstrating his grasp of the forefront of technology, wise,confident,anime style --ar 1:1 --niji 5

CEO+profound

提示词：male, the CEO stands in front of the fashion car exhibition booth, profound, his gaze insightful of market trends, his gestures powerfully pointing towards dazzling cars, reflecting his understanding of the market, anime style --ar 1:1 --niji 5

CEO+profound

提示词：male, close shot,the CEO walked on the landing pad of a private helicopter, his gaze fixed on the distance,his documents seemed to hold onto the commercial future,profound, anime style --ar 1:1 --niji 5

第2章

俊美男

俊美男的面容精致，使人不由自主地被吸引住。他们气质优雅，举止文雅，好像被赋予了一种与生俱来的高贵魅力。无论是正式场合还是休闲时光，俊美男总是能穿搭得体，为自己赋予一种独特的气质。

关键词：handsome man, fashionable, confident, approachable, eye-catching, caring, charming, confidence

confident+fashionable

提示词：stylish man, vibrant floral shirt, tousled hair, confident expression,fashionable, anime style --ar 1:1 --niji 5

handsome man+approachable

提示词：handsome man, textured hair, approachable, anime style --ar 1:1 --niji 5

confident+handsome man

提示词：handsome man, the fashionable attire of the flower beauty man complements the artwork, confident，becoming a unique landscape in the city, anime style --ar 1:1 --niji 5

handsome man+charming

提示词：handsome man, charming,his elegant demeanor and fashionable attire attract the attention of people around him, as if it were a beautiful scenery line,anime style --ar 1:1 --niji 5

handsome man+confident

提示词：handsome man, walking on the golden beach, confident, the figure and skin color of the flower beauty man complement each other, attracting all the attention in the sunlight, anime style --ar 1:1 --niji 5

handsome man+confident

提示词：handsome man, attracted a passionate audience with the fashionable design and energetic dance movements of the flower beauty man, confident,anime style --ar 1:1 --niji 5

handsome man+charming

提示词：handsome man, gracefully savors with the beautiful flowers, his presence makes the entire environment even more charming,anime style --ar 1:1 --niji 5

handsome man+approachable

提示词：handsome man, on the grass in the picnic garden, shares food and laughter with his friends, his approachable make this outdoor gathering even more warm, anime style --ar 1:1 --niji 5

handsome man+eye-catching

提示词：handsome man, the taste and temperament of his complement his artistic works,eye-catching, he is also a beautiful scenery in the atmosphere of art and beauty, anime style --ar 1:1 --niji 5

handsome man+charming

提示词：handsome man, lawn of a park concert, admires the music while swaying to the rhythm, charming, his presence adds a charming charm to the concert, nime style --ar 1:1 --niji 5

handsome man+charming

提示词：handsome man, the outstanding outfits of the flower beauty man complement the natural scenery, charming，as if it were an intoxicating scene, anime style --ar 1:1 --niji 5

handsome man+charming

提示词：mid-shot,handsome man, rowing or surfing in the lake, the water sports skills of his complement his sunny smile,charming，anime style --ar 1:1 --niji 5

handsome man+charming

提示词：handsome man,on the snow-white ski resort,the flower beauty man showcases his athletic talent, as he glides through the snow, he is full of challenge spirit, charming, anime style --ar 1:1 --niji 5

handsome man+charming

提示词：handsome man,the flower beauty man blends with nature,charming， his appearance and inner qualities seem to be a gift from nature, anime style --ar 1:1 --niji 5

handsome man+charming

提示词：mid shot, handsome man, the flower beauty man blends with nature, charming, his appearance and inner qualities seem to be a gift from nature, anime style --ar 1:1 --niji 5

handsome man+confidence

提示词：mid shot, handsome man, showcasing vitality and confidence and firm steps of the beautiful man, anime style --ar 1:1 --niji 5

handsome man+charming

提示词：mid shot, handsome man, sitting on a lounge chair in the open-air cinema at night, while the beautiful man enjoys the movie, his charming emits a charming light under the starry sky, anime style --ar 1:1 --niji 5

handsome man+caring

提示词：handsome man, caring, skilled in assembling tents in a wilderness camping site, showcases his adventurous spirit and leadership,becoming the soul figure of the team, anime style --ar 1:1 --niji 5

handsome man+charming

提示词：handsome man,the performance is full of vitality and passion, his dance moves and music blend together,charming， attracting cheers from the audience, anime style --ar 1:1 --niji 5

handsome man+charming

提示词：handsome man, records the beauty of nature with his brush, charming, his artistic talent and focus become a touching scenery, anime style --ar 1:1 --niji 5

handsome man+charming

提示词：handsome man, strolling in the pastoral shooting scene, the temperament of the flower beauty man blends with the natural environment,charming, his presence illuminates the entire scene, anime style --ar 1:1 --niji 5

handsome man+confident

提示词：handsome man,confident,the resolute expression of the magnificent scenery the background complement each other, anime style --ar 1:1 --niji 5

handsome man+approachable

提示词：handsome man,demonstrating practical abilities hardworking spirit,approachable, his versatility is admirable, anime style --ar 1:1 --niji 5

handsome man+charming

提示词：handsome man, in an outdoor studio, charming, his natural expression and versatile design become the highlight in front of the camera, anime style --ar 1:1 --niji 5

handsome man+confident

提示词：handsome man, confident, sitting by the lake fishing, the tranquility and patience make his catch not only fish, but also a symbol of tranquility and satisfaction, anime style --ar 1:1 --niji 5

handsome man+approachable

提示词：handsome man, in the orchard of the farmland,approachable, his clever gestures and cheerful smile add a beautiful life to the rural scenery, anime style --ar 1:1 --niji 5

handsome man+eye-catching

提示词：handsome man, eye-catching, accompanied by various flowers, his elegant temperament complements the harmony of nature, anime style --ar 1:1 --niji 5

handsome man+fashionable

提示词：handsome man,adds a sense of fashion and liveliness to the market with the fashionable attire and friendly smile of the flower beauty man, anime style --ar 1:1 --niji 5

handsome man+approachable

提示词：handsome man, sitting in the circle of an outdoor bonfire party, approachable，his musical talent and affinity bloomed in the bonfire at night, anime style --ar 1:1 --niji 5

handsome man+charming

提示词：handsome man,standing the beach at sunset,with a gentle breeze and golden sunlight shining on his perfect profile，charming, anime style --ar 1:1 --niji 5

handsome man+eye-catching

提示词：handsome man,riding a camel across the desert,showcases the courage and adventurous spirit of a beautiful man in the vast desert, eye-catching, anime style --ar 1:1 --niji 5

handsome man+charming

提示词：handsome man, the relaxed expression, charming，making his relaxation and comfort a part of the attraction, anime style --ar 1:1 --niji 5

handsome man+charming

提示词：handsome man, in the corner of the library, dedicated to reading, charming, his deep eyes revealing wisdom and contemplation, anime style --ar 1:1 --niji 5

handsome man+fashionable

提示词：handsome man,fashionable， his hair was wet, his eyes showed a playful expression that made people move endlessly, anime style --ar 3:4

handsome man+charming

提示词:handsome man,in the concert hall,he played the piano,notes jumping at his fingertips, charming and melodious melodies filling the air, anime style --ar 1:1 --niji 5

handsome man+eye-catching

提示词：handsome man,eye-catching, exudes tranquility through his smile, anime style --ar 1:1 --niji 5

handsome man+charming

提示词：handsome man,in the snow, gently raised snowflakes,charming，his smile filled with pure joy, anime style --ar 1:1 --niji 5

handsome man+eye-catching

提示词：handsome man,eye-catching，gliding on the ice with a relaxed posture like a dancer on stage, anime style --ar 1:1 --niji 5

第3章

优雅绅士

优雅绅士类型的美男子是受人尊敬和令人钦佩的，他们具备较高的文化修养，通常在文学、艺术、历史、哲学等方面拥有较深的造诣。他们还具有高尚的品德，仁爱、正直、宽容、谦虚。

关键词：charming and refined gentleman, gentleman, elegant, calm, dressed, confidence, meticulousness, sensitivity, attentively, admiring, captivating, confidently, thoughtful, affinity, emotional, meticulously, humorous, peaceful, scientific, talent, sensitivity, athletic, joy, patience, natural charm, leisurely,adventurous, peaceful

charming and refined gentleman+confidence

提示词：charming and refined gentleman, exuding confidence, captivating smile, anime style --ar 1:1 --niji 5

gentleman+dressed

提示词：suave and sophisticated gentleman, dressed in a tailored suit, charming demeanor, captivating smile, anime style --ar 1:1 --niji 5

charming and refined gentleman+elegant

提示词：charming and refined gentleman, who speaks appropriately at social gatherings and wins people's attention with elegant demeanor, anime style --ar 1:1 --niji 5

charming and refined gentleman

提示词：charming and refined gentleman, a gentleman demonstrates vitality, as well as respect and care for nature, anime style --ar 1:1 --niji 5

charming and refined gentleman+confidently

提示词：charming and refined gentleman, a gentleman who focuses and confidently completes tasks, demonstrating leadership and sense of responsibility with an efficient and professional attitude, anime style --ar 1:1 --niji 5

charming and refined gentleman+meticulous

提示词：charming and refined gentleman, a gentleman interacts with people and demonstrates meticulous etiquette at a dinner party, anime style --ar 1:1 --niji 5

charming and refined gentleman+sensitivity

提示词：charming and refined gentleman, a gentleman appreciates culture at art exhibitions or concerts, demonstrating sensitivity and taste towards art, anime style --ar 1:1 --niji 5

charming and refined gentleman+calm

提示词：charming and refined gentleman, maintaining calm during the journey and demonstrating a spirit of exploration, anime style --ar 1:1 --niji 5

charming and refined gentleman+attentively

提示词：charming and refined gentleman, a gentleman smiled at the concert, listening attentively to the wonderful music, anime style --ar 1:1 --niji 5

charming and refined gentleman

提示词：charming and refined gentleman, in the rain, the gentleman opens his umbrella, anime style --ar 1:1 --niji 5

charming and refined gentleman+confidence

提示词：charming and refined gentleman, at business meetings, earned the respect of his peers for their eloquence, confidence, humility, anime style --ar 1:1 --niji 5

charming and refined gentleman+admiring

提示词：charming and refined gentleman, a gentleman dressed appropriately, stood in front of the gallery, admiring artworks, with a unique appreciation of beauty in his eyes, anime style --ar 1:1 --niji 5

charming and refined gentleman

提示词：charming and refined gentleman, on a yacht, his smile exudes a love for ocean freedom, anime style --ar 1:1 --niji 5

charming and refined gentleman+captivating

提示词：charming and refined gentleman, the gentleman sat in the corner of the coffee shop, flipping through ancient books, his deep thinking and introversion captivating, anime style --ar 1:1 --niji 5

charming and refined gentleman+meticulousness

提示词：charming and refined gentleman, showcasing their meticulousness and thoughtfulness, anime style --ar 1:1 --niji 5

charming and refined gentleman+confidence

提示词：charming and refined gentleman, dressed in meticulously matched suits, exudes fluency and confidence in public speeches, attracting the attention of the audience, anime style --ar 1:1 --niji 5

charming and refined gentleman+thoughtful

提示词：charming and refined gentleman, in front of the chess game, the gentleman pondered his next move, revealing calm and thoughtful movements, anime style --ar 1:1 --niji 5

charming and refined gentleman+affinity

提示词：charming and refined gentleman,strolling through the park, showcasing his respect and affinity, anime style --ar 1:1 --niji 5

charming and refined gentleman+elegant

提示词：charming and refined gentleman, the gentleman sat in the corner of the café, quietly reading the newspaper, with elegant demeanor, anime style --ar 1:1 --niji 5

charming and refined gentleman

提示词：charming and refined gentleman, standing next to a classic car on the city street, exudes his demeanor, anime style --ar 1:1 --niji 5

charming and refined gentleman+emotional

提示词：charming and refined gentleman, the gentleman sat in front of the piano, emitting emotional notes, anime style --ar 1:1 --niji 5

charming and refined gentleman+meticulously

提示词：charming and refined gentleman,the gentleman meticulously carves handicrafts, showcasing his patience and craftsmanship at every step, anime style --ar 1:1 --niji 5

charming and refined gentleman+humorous

提示词：charming and refined gentleman, in the bar, chat easily with friends, revealing a humorous and witty side, anime style --ar 1:1 --niji 5

charming and refined gentleman+peaceful

提示词：charming and refined gentleman, the gentleman sits on comfortable sofa, creating a peaceful atmosphere, anime style --ar 1:1 --niji 5

charming and refined gentleman+scientific

提示词：charming and refined gentleman, a gentleman at a technology, explained the principles of high-tech products in simple and profound terms, presenting his scientific knowledge and charm, anime style --ar 1:1 --niji 5

charming and refined gentleman+talent

提示词：charming and refined gentleman, in the studio,the gentleman meticulously creates his paintings, each stroke and color showcasing his artistic talent, anime style --ar 1:1 --niji 5

charming and refined gentleman

提示词：charming and refined gentleman, a gentleman in the library, with a longing and respect for knowledge in his eyes, anime style --ar 1:1 --niji 5

charming and refined gentleman+sensitivity

提示词：charming and refined gentleman, the gentleman walks past clusters of flowers, demonstrating his sensitivity and love for nature, anime style --ar 1:1 --niji 5

charming and refined gentleman+athletic

提示词：charming and refined gentleman, the gentleman swings and hits the ball on the golf course, showcasing his athletic style and elegant leisure style, anime style --ar 1:1 --niji 5

charming and refined gentleman+joy

提示词：charming and refined gentleman,at outdoor concerts, the gentleman enjoys music and enjoys the joy of music together with people, anime style --ar 1:1 --niji 5

charming and refined gentleman

提示词：charming and refined gentleman,walking with dogs in a suburban park, anime style --ar 1:1 --niji 5

charming and refined gentleman+patience

提示词：charming and refined gentleman, a gentleman fishing by the river, waiting for the fish to catch the bait, demonstrating his patience and relaxed approach, anime style --ar 1:1 --niji 5

charming and refined gentleman

提示词：charming and refined gentleman, showcasing his pursuit of quality living at an outdoor open-air market, anime style --ar 1:1 --niji 5

charming and refined gentleman

提示词：charming and refined gentleman, dressed in a suit, stood in front of the flower bed in the city park, admiring the colorful flowers, anime style --ar 1:1 --niji 5

charming and refined gentleman+natural charm

提示词：charming and refined gentleman, at a rural estate, the gentleman showcases his natural charm and tacit understanding with animals, anime style --ar 1:1 --niji 5

charming and refined gentleman+leisurely

提示词：charming and refined gentleman, the gentleman wearing a sunshade hat, leisurely strolls on the beach, showcasing his leisure and vacation style, anime style --ar 1:1 --niji 5

charming and refined gentleman+adventurous

提示词：charming and refined gentleman, a gentleman who travels through forest paths, backpacks on foot to explore, close to nature, showcasing his adventurous side, anime style --ar 1:1 --niji 5

charming and refined gentleman+peaceful

提示词：charming and refined gentleman, a gentleman wearing a coat on the beach pier, revealing his profound thinking and peaceful heart, anime style --ar 1:1 --niji 5

第4章

古风美男

古风人物以外表、品德、修养和行为展现出了中国古代文化的独特魅力。古风人物的外貌和服饰往往是引人注目的地方，这些打扮和穿着也反映出了那个时代的背景和人物所处的社会地位。他们对待生活和事业充满热情，勇于面对困难和挑战，坚持自己的信仰和原则，这种坚韧和决心使他们能够克服各种困难，取得成功。古风人物还非常注重自身的修养，他们热爱文学、书法、绘画和音乐，饱读诗书，以此培养内在的高尚品质。

关键词：elegant and graceful ancient, handsome, handsome man in traditional costume

handsome

提示词：flowing robes, confident and charismatic expression, handsome man， anime style--ar 1:1 --niji 5

handsome man in traditional costume

提示词：handsome man in traditional costume, sharp and piercing eyes, anime style--ar 1:1 --niji 5

elegant and graceful ancient handsome

提示词：elegant and graceful ancient handsome man, stepping on the wind and coming, his clothes flowing, anime style--ar 1:1 --niji 5

elegant and graceful ancient handsome

提示词：elegant and graceful ancient handsome man, strolling along the bluestone path in front of the ancient temple, his gaze deep, anime style--ar 1:1 --niji 5

elegant and graceful ancient handsome

提示词：elegant and graceful ancient handsome man,standing by the riverbank, his long hair swaying gently in the gentle breeze, picturesque and poetic, anime style--ar 1:1 --niji 5

elegant and graceful ancient handsome

提示词：elegant and graceful ancient handsome man, dressed in ancient attire, strolls in the courtyard of plum blossoms, his smile brushing his face like spring breeze, anime style--ar 1:1 --niji 5

elegant and graceful ancient handsome

提示词：elegant and graceful ancient handsome man, standing deep in the bamboo forest, attracted the birds to stop,anime style--ar 1:1 --niji 5

elegant and graceful ancient handsome

提示词：elegant and graceful ancient handsome man,standing on an ancient bridge, looking into the distance, his gaze contains a indifference to the flow of time, anime style--ar 1:1 --niji 5

elegant and graceful ancient handsome

提示词：elegant and graceful ancient handsome man, a white horse, crossing the ancient road, his gaze steadfast, anime style--ar 1:1 --niji 5

elegant and graceful ancient handsome

提示词：elegant and graceful ancient handsome man,standing on the mountaintop, his gaze was like a bright moon, distant and cold, anime style--ar 1:1 --niji 5

elegant and graceful ancient handsome

提示词：elegant and graceful ancient handsome man, strolling on the grassland, his clothes fluttering in the breeze,anime style--ar 1:1 --niji 5

elegant and graceful ancient handsome

提示词：elegant and graceful ancient handsome man, standing on the ancient city, he gazes into the distance, anime style--ar 1:1 --niji 5

elegant and graceful ancient handsome

提示词：elegant and graceful ancient handsome man,among the green mountains and waters, he holds a folding fan and smiles like a spring breeze, anime style--ar 1:1 --niji 5

elegant and graceful ancient handsome

提示词：elegant and graceful ancient handsome man,walking in the ancient streets and alleys, his gaze like stars, reveals a deep affection for history, anime style--ar 1:1 --niji 5

elegant and graceful ancient handsome

提示词：elegant and graceful ancient handsome man, standing at the riverbank, his hair brushed by the gentle breeze, anime style--ar 1:1 --niji 5

elegant and graceful ancient handsome

提示词：elegant and graceful ancient handsome man, crossing the sea of flowers, he holds the flowers in his hand, his smile is warm as sunshine, anime style--ar 1:1 --niji 5

elegant and graceful ancient handsome

提示词：elegant and graceful ancient handsome man, dressed in ancient attire, strolls along the shaded mountain path, like a fairy descending to earth, anime style--ar 1:1 --niji 5

elegant and graceful ancient handsome

提示词：elegant and graceful ancient handsome man,standing on a high platform,overlooking the earth, exuding a regal aura, anime style--ar 1:1 --niji 5

elegant and graceful ancient handsome

提示词：elegant and graceful ancient handsome man,strolling by the lake, reflecting on the lake surface, anime style--ar 1:1 --niji 5

elegant and graceful ancient handsome

提示词：elegant and graceful ancient handsome man, leading a horse, his posture is elegant and composed, anime style--ar 3:4 --niji 5

elegant and graceful ancient handsome

提示词：elegant and graceful ancient handsome man, in the garden, anime style--ar 1:1 --niji 5

elegant and graceful ancient handsome

提示词：elegant and graceful ancient handsome man, stopping by the waterfall, anime style--ar 1:1 --niji 5

elegant and graceful ancient handsome

提示词：elegant and graceful ancient handsome man, crossing the forest path, his smile bright as sunlight, scattering on the ground,anime style--ar 1:1 --niji 5

elegant and graceful ancient handsome

提示词：elegant and graceful ancient handsome man, stepping on the beach, his steps are light and free, anime style--ar 1:1 --niji 5

elegant and graceful ancient handsome

提示词：elegant and graceful ancient handsome man, strolling in the autumn forest, his eyes as clear as autumn water, anime style--ar 1:1 --niji 5

elegant and graceful ancient handsome

提 示 词：elegant and graceful ancient handsome man, walking slowly between mountains and rivers, dressed in an antique robe, anime style--ar 1:1 --niji 5

elegant and graceful ancient handsome

提示词：elegant and graceful ancient handsome man, sitting quietly in the courtyard, his breath calm and restrained, anime style--ar 1:1 --niji 5

elegant and graceful ancient handsome

提示词：elegant and graceful ancient handsome man, walking on a rural path, holding a bouquet of wild flowers in his hand, exuding a simple and pure beauty, anime style--ar 1:1 --niji 5

elegant and graceful ancient handsome

提示词：elegant and graceful ancient handsome man, lying on the bluestone by the lake, he closed his eyes and enjoyed the cool breeze, anime style--ar 1:1 --niji 5

elegant and graceful ancient handsome

提示词：elegant and gracefu ancient handsome man, sitting by the window, holding a guqin and gently playing it, anime style--ar 1:1 --niji 5

elegant and graceful ancient handsome

提示词：elegant and graceful ancient handsome man, in the study,he lowered his head in contemplation, anime style--ar 1:1 --niji 5

elegant and graceful ancient handsome

提示词：elegant and graceful ancient handsome man, under the emerald curtains, he holds a jade flute in his hand and plays melodious melodies, anime style--ar 1:1 --niji 5

elegant and graceful ancient handsome

提示词：elegant and graceful ancient handsome man, standing under the candlelight, anime style--ar 1:1 --niji 5

elegant and graceful ancient handsome

提示词：elegant and graceful ancient handsome man, sitting on the windowsill, caressing plum blossoms, anime style--ar 1:1 --niji 5

elegant and graceful ancient handsome

提示词：elegant and graceful ancient handsome man, sitting on a chair, anime style--ar 1:1 --niji 5

elegant and graceful ancient handsome

提示词：elegant and graceful ancient handsome man, standing in the courtyard admiring flowers, anime style--ar 1:1 --niji 5

elegant and graceful ancient handsome

提示词：elegant and graceful ancient handsome man, standing among the mountains and rivers, dressed in a simple robe, the mountains and waters, anime style--ar 1:1 --niji 5

elegant and graceful ancient handsome

提示词：elegant and graceful ancient handsome man, sitting next to the tea stand, sipping tea as the breeze blew, anime style--ar 1:1 --niji 5

第5章

美少年

美少年总是充满活力的，给人们无限的希望。他们招人喜爱，吸引力在他们身上体现得淋漓尽致。他们在面对挑战和困难时表现出坚定的意志和勇气，这种决心和勇敢也是他们的魅力。同时他们纯真且善良，不禁让人想要保护他们！

关键词： beautiful boy full of life and vitality, elegance, bright smile, charming, youthful exuberance, contemplates, gentle

beautiful boy full of life and vitality

提示词：beautiful boy full of life and vitality, thick hair, bright and sunny，handsome man，anime style--ar 1:1 --niji 5

charming+youthful exuberance

提示词：a beautiful young man who gives people infinite hope,charming,hick hair,youthful exuberance, handsome man, anime style--ar 1:1 --niji 5

beautiful boy full of life and vitality

提示词： beautiful boy full of life and vitality, standing in the garden in the morning light, his beautiful hair fluttering with the breeze, anime style--ar 1:1 --niji 5

beautiful boy full of life and vitality

提示词 : at dusk, beautiful boy full of life and vitality, with his back to the sun,anime style--ar 3:4 --niji 5

beautiful boy full of life and vitality

提示词：beautiful boy full of life and vitality,with the sea breeze brushing his hair, anime style--ar 1:1 --niji 5

beautiful boy full of life and vitality

提示词：beautiful boy full of life and vitality, standing in the green bamboo forest, as if integrated with nature, anime style--ar 1:1 --niji 5

beautiful boy full of life and vitality

提示词：beautiful boy full of life and vitality, on the boulevard, the beautiful boy carries a backpack, his smile looks particularly bright in the sunlight, anime style--ar 1:1 --niji 5

beautiful boy full of life and vitality

提示词：beautiful boy full of life and vitality, in the valley, the beautiful boy rides a white horse, anime style--ar 1:1 --niji 5

beautiful boy full of life and vitality

提示词：beautiful boy full of life and vitality, on the grassland, the beautiful boy holds flowers and faces the vast sky, anime style--ar 1:1 --niji 5

beautiful boy full of life and vitality

提示词：beautiful boy full of life and vitality, the beautiful boy devoutly offers his prayers, anime style--ar 1:1 --niji 5

beautiful boy full of life and vitality

提示词：beautiful boy full of life and vitality, on the top of a snowy mountain, looking into the distance, his eyes are deep and peaceful, anime style--ar 1:1 --niji 5

beautiful boy full of life and vitality

提示词：beautiful boy full of life and vitality, on the quiet lakeside, his smile swept over the lake like a spring breeze, rippling, anime style--ar 1:1 --niji 5

beautiful boy full of life and vitality

提示词：beautiful boy full of life and vitality, strolling through ancient streets and alleys, his eyes filled with curiosity and respect for history, anime style--ar 1:1 --niji 5

beautiful boy full of life and vitality

提示词：beautiful boy full of life and vitality,the gentle breeze caresses his hair, his smile spreads like sunlight on the earth, anime style--ar 1:1 --niji 5

beautiful boy full of life and vitality

提示词：beautiful boy full of life and vitality, lying on the grass under the starry sky, looking up at the twinkling stars, anime style--ar 1:1 --niji 5

beautiful boy full of life and vitality

提示词：beautiful boy full of life and vitality, in the morning light, the beautiful boy holds a camera and focuses on capturing the beautiful scenery of nature, anime style--ar 1:1 --niji 5

beautiful boy full of life and vitality

提示词：beautiful boy full of life and vitality, in the desert town, leading camels,anime style--ar 3:4 --niji 5

beautiful boy full of life and vitality+bright smile

提示词：beautiful boy full of life and vitality, in the forest, with a bright smile like sunshine, anime style--ar 1:1 --niji 5

beautiful boy full of life and vitality+elegance

提示词：beautiful boy full of life and vitality, on the ice covered lake, the beautiful boy is skating, full of elegance and agility, anime style--ar 1:1 --niji 5

beautiful boy full of life and vitality

提示词：beautiful boy full of life and vitality, sitting under a tree in autumn with fallen leaves, holding a book in hand, anime style--ar 1:1 --niji 5

beautiful boy full of life and vitality

提示词：beautiful boy full of life and vitality, under the sunlight, the beautiful boy dressed in loose clothes, holding a bamboo flute, anime style--ar 1:1 --niji 5

beautiful boy full of life and vitality

提示词：beautiful boy full of life and vitality, as night fell on the rooftop, behind the starry night sky, his eyes filled with mysterious depth， anime style--ar 1:1 --niji 5

beautiful boy full of life and vitality+bright

提示词：beautiful boy full of life and vitality,water droplets splashing beside him, his hair moist, but his smile still bright, anime style--ar 1:1 --niji 5

beautiful boy full of life and vitality

提示词：beautiful boy full of life and vitality, sitting in a rocking chair by the lake at dusk, his mood peaceful and willing, anime style--ar 1:1 --niji 5

beautiful boy full of life and vitality

提示词：beautiful boy full of life and vitality, walking in the sea of cherry blossoms, his eyes filled with longing for beauty, anime style--ar 1:1 --niji 5

beautiful boy full of life and vitality+bright smile

提示词：beautiful boy full of life and vitality, walking on the beach, facing the sea breeze,bright smile, anime style--ar 1:1 --niji 5

beautiful boy full of life and vitality

提示词：beautiful boy full of life and vitality, walking with firm steps, his face filled with a desire for exploration, anime style--ar 1:1 --niji 5

beautiful boy full of life and vitality

提示词：beautiful boy full of life and vitality, at an outdoor concert,the beautiful boy listens to the music, anime style--ar 1:1 --niji 5

beautiful boy full of life and vitality+contemplates

提示词：beautiful boy full of life and vitality, in the exquisite living room, the beautiful boy gracefully savors a cup of black tea and contemplates life, anime style--ar 1:1 --niji 5

beautiful boy full of life and vitality

提示词：beautiful boy full of life and vitality, playing the piano in the music room, filled with artistic emotions, anime style--ar 1:1 --niji 5

beautiful boy full of life and vitality

提 示 词：beautiful boy full of life and vitality, in the studio, the beautiful boy holds a paintbrush and focuses on outlining the scenery on the canvas, his creations are full of inspiration and emotion, anime style--ar 1:1 --niji 5

beautiful boy full of life and vitality

提示词：beautiful boy full of life and vitality, the beautiful boy is preparing a delicate dinner with his own hands, his smile is as warm as sunshine, anime style--ar 1:1 --niji 5

beautiful boy full of life and vitality

提示词：beautiful boy full of life and vitality, in the dance room, the beautiful boy performed a graceful dance, his movements light and full of passion, anime style--ar 1:1 --niji 5

beautiful boy full of life and vitality

提示词：beautiful boy full of life and vitality, the beautiful boy is immersed in the ocean of artistic works, his eyes exude praise for art, anime style--ar 1:1 --niji 5

beautiful boy full of life and vitality

提示词：beautiful boy full of life and vitality, in the control room of modern technology, the beautiful boy is wearing VR devices and immersed in the virtual world, his expression filled with curiosity, anime style--ar 1:1 --niji 5

beautiful boy full of life and vitality+gentle

提示词：beautiful boy full of life and vitality, standing on the balcony, facing the first ray of sunshine in the morning, gentle and peaceful, anime style--ar 1:1 --niji 5

beautiful boy full of life and vitality

提示词：beautiful boy full of life and vitality, in the garden in the rain, the beautiful boy wears his coat and admires the scene of fresh rain washing the earth, his mood is as bright as a flower after the rain, anime style--ar 1:1 --niji 5

beautiful boy full of life and vitality

提示词：beautiful boy full of life and vitality, strolling in the bustling crowd on the city streets with the sunset, his smile warming people's hearts like the setting sun, anime style--ar 1:1 --niji 5

第6章

魅力大叔

大叔经历过岁月的洗礼，充满了阅历和智慧。他们的魅力在于深沉的内在、优雅的品质以及对生活的理解。他们有敏锐的洞察力，在处理复杂的人际关系和情感问题时游刃有余。他们的言行也常常能激起人们的深思和共鸣。

关键词：handsome, middle-aged man, mature and steady, considerate, humorous and witty, knowledgeable, patient, firm, confident, tasteful, gracefully, humor

handsome+middle-aged man+mature and steady

提示词：handsome, middle-aged man, mature and steady, round glasses, short hair, anime style--ar 1:1 --niji 5

handsome+middle-aged man+tasteful

提示词：handsome, middle-aged man with a large back combed head, tasteful, white shirt, anime style--ar 1:1 --niji 5

handsome+middle-aged man+gracefully

提示词：handsome, middle-aged man, gracefully savoring red wine in the living room, anime style--ar 1:1 --niji 5

handsome+middle-aged man

提示词：handsome, middle-aged man, fine and dense beard,sitting at a desk, focused on reading classic literary works, anime style--ar 1:1 --niji 5

handsome+middle-aged man

提示词：handsome, middle-aged man, fine and dense beard, in the kitchen, skillfully cooking, delicious dishes were born in his hands, anime style--ar 1:1 --niji 5

handsome+middle-aged man+knowledgeable

提示词：handsome, middle-aged man, knowledgeable, anime style--ar 1:1 --niji 5

handsome+middle-aged man+patient

提示词：handsome, middle-aged man, patient, fine and dense beard, in the home theater, watching classic movies, his smile was filled with warmth and happiness, anime style--ar 1:1 --niji 5

handsome+middle-aged man+confident

提示词：handsome, middle-aged man, sitting, confident, demonstrating his leadership skills, colleagues deeply respect his opinions, anime style--ar 1:1 --niji 5

handsome+middle-aged man

提示词：handsome, middle-aged man, fine and dense beard, performing fitness exercises in the gym demonstrates his strong willpower and healthy attitude towards life, anime style--ar 1:1 --niji 5

handsome+middle-aged man+humorous and witty

提示词：handsome, middle-aged man, humorous and witty, the charming uncle is accompanied by a book, quietly enjoying a peaceful afternoon, anime style--ar 1:1 --niji 5

handsome+middle-aged man

提示词：handsome, middle-aged man, fine and dense beard, in the indoor garden, the charming uncle carefully takes care of various flowers and plants, anime style--ar 1:1 --niji 5

handsome+middle-aged man+humor

提示词：handsome, middle-aged man,chatting freely about life, sharing his wisdom and sense of humor, anime style--ar 1:1 --niji 5

handsome+middle-aged man+mature and steady

提示词：handsome, middle-aged man, mature and steady,the charming uncle interweaves with the gentle breeze and waves to create a peaceful and elegant scene, anime style--ar 1:1 --niji 5

handsome+middle-aged man+considerate

提示词：handsome, middle-aged man, considerate, short hair, take a walk with dog in the park, anime style--ar 1:1 --niji 5

handsome+middle-aged man+confident

提示词：handsome, middle-aged man, confident, fine and dense beard, at the outdoor coffee shop, the charming uncle savors the rich coffee, anime style--ar 1:1 --niji 5

handsome+middle-aged man+firm

提示词：handsome, middle-aged man, firm, charming uncle adheres to a healthy lifestyle and enjoys the beautiful scenery of nature, anime style--ar 1:1 --niji 5

handsome+middle-aged man+mature and steady

提示词：handsome, middle-aged man, fine and dense beard, mature and steady, the charming uncle demonstrates his diligence and responsibility, anime style--ar 1:1 --niji 5

handsome+middle-aged man+tasteful

提示词：handsome, middle-aged man, at the outdoor music festival, enthusiasm attracts people around, tasteful, anime style--ar 1:1 --niji 5

handsome+middle-aged man+firm

提示词：handsome, middle-aged man, firm, on the football field, cheer for the team, anime style--ar 1:1 --niji 5

handsome+middle-aged man+humorous and witty

提示词：handsome, middle-aged man, humorous and witty, a smile permeates the harmonious atmosphere, anime style--ar 1:1 --niji 5

handsome+middle-aged man+confident

提示词：handsome, middle-aged man, confident, fine and dense beard, in the outdoor courtyard, lights a bonfire, anime style--ar 1:1 --niji 5

handsome+middle-aged man+knowledgeable

提示词：handsome, middle-aged man, knowledgeable, enjoying skiing at a ski resort, he is skilled in his skills, anime style--ar 1:1 --niji 5

handsome+middle-aged man+firm

提示词：handsome, middle-aged man, firm, uncle stood on the peak of a high mountain, facing the wind and challenging, anime style--ar 1:1 --niji 5

handsome+middle-aged man+mature and steady

提示词：handsome, middle-aged man, mature and steady, he was fishing on the dock by the sea, waiting quietly, anime style--ar 1:1 --niji 5

handsome+middle-aged man+mature and steady

提示词：handsome, middle-aged man, mature and steady, fine and dense beard, sitting on the park bench, waiting quietly, anime style--ar 1:1 --niji 5

handsome+middle-aged man+patient

提示词：handsome, middle-aged man, patient, tracksuit, runs on the street of the city, anime style--ar 1:1 --niji 5

handsome+middle-aged man+knowledgeable

提示词：a middle-aged man, knowledgeable, handsome man, fine and dense beard, in the manor, anime style--ar 3:4 --niji 5

handsome+middle-aged man+confident

提示词：handsome, middle-aged man, confident, fine and dense beard, driving a boat, mastering the direction, anime style--ar 1:1 --niji 5

handsome+middle-aged man+humorous and witty

提示词：handsome, middle-aged man, humorous and witty, fine and dense beard, he was reading by the fire in the mountain cottage, immersed in books, anime style--ar 1:1 --niji 5

handsome+middle-aged man+confident

提示词：handsome, middle-aged man, confident, playing the guitar at the outdoor music festival, anime style--ar 1:1 --niji 5

handsome+middle-aged man+knowledgeable

提示词：handsome, middle-aged man, knowledgeable, fine and dense beard, standing on the sunny mountain peak, his resilience and determination are admirable, anime style--ar 1:1 --niji 5

handsome+middle-aged man+confident

提示词：handsome, middle-aged man, confident, takes a walk on the street of the city, anime style--ar 1:1 --niji 5

handsome+middle-aged man+mature and steady

提示词：handsome, middle-aged man, mature and steady, fine and dense beard, lying on the grass, anime style--ar 1:1 --niji 5

handsome+middle-aged man+tasteful

提示词：handsome, middle-aged man, white shirt, short hair, walks in the forest, tasteful, anime style--ar 1:1 --niji 5

handsome+middle-aged man+and tasteful

提示词：handsome, middle-aged man, tasteful, fine and dense beard, sitting outside the tent, anime style--ar 1:1 --niji 5

handsome+middle-aged man+knowledgeable

提示词：handsome, middle-aged man, knowledgeable, fine and dense beard, stands in front of the castle, anime style--ar 1:1 --niji 5

handsome+middle-aged man+patient

提示词：handsome, middle-aged man, patient, fine and dense beard, wearing professional climbing equipment, climbing with a firm and focused gaze, anime style--ar 1:1 --niji 5

handsome+middle-aged man+mature and steady

提示词：handsome, middle-aged man, mature and steady, holding a cup of hot chocolate in his hand, admiring the pure white snow, anime style--ar 1:1 --niji 5

第7章

艺术家

艺术家才华横溢，总是富有创造力，他们通过各种艺术形式表达内心世界和情感，他们的魅力源于才华、情感表达和对美的追求。他们的作品充满了浓厚的情感，使人们产生共鸣。艺术家还常常追求美的极致，他们在作品中探索美的各种形式和表现，让观众得到美的启发。他们从不畏惧表达自己的观点和声音，通过作品传递出真实的自我。

关键词：a handsome art designer, fashionable, elegant, confident, creative, free, exquisite

a handsome art designer+elegant

提示词：male, a handsome art designer, elegant, wearing personality, very charming, anime style--ar 1:1 --niji 5

a handsome art designer+elegant

提示词：male, a handsome art designer, dressing with personality, elegant, wearing black-rimmed glasses, anime style--ar 1:1 --niji 5

a handsome art designer+creative

提示词：male, a handsome art designer, creative,sitting on a bench in a city park,with blooming flowers, immersed in capturing the greenery of the park and urban life, anime style--ar 1:1 --niji 5

a handsome art designer+exquisite

提示词：male, a handsome art designer, exquisite, with sunlight shining on his shoulders, anime style--ar 1:1 --niji 5

a handsome art designer+fashionable

提示词：male, a handsome art designer, on the beach by the sea, fashionable， the artist stands in front of the easel, forever capturing the tranquility and beauty of this moment on the canvas, anime style--ar 1:1 --niji 5

a handsome art designer+fashionable

提示词：male, a handsome art designer, fashionable, with the canvas resting on his knees, he lightly touches it with his pen tip, anime style--ar 1:1 --niji 5

a handsome art designer+fashionable

提示词：male, fashionable, a handsome art designer,in the countryside in the suburbs, the artist lies on the green grass, anime style--ar 1:1 --niji 5

a handsome art designer+fashionable

提示词：male, a handsome art designer, is located in a city center where the artist sits in front of a small stall and draws paintings, fashionable, anime style--ar 1:1 --niji 5

a handsome art designer+free

提示词：a handsome art designer, male, free, deep in the rainforest, the artist sits under a tree and uses a brush to depict the work, anime style--ar 1:1 --niji 5

a handsome art designer+free

提示词：a handsome art designer, male, free, draws paintings, preserves the beauty of nature forever in his works, anime style--ar 1:1 --niji 5

a handsome art designer+free

提示词：a handsome art designer, free, male, artist stands on high mountain, peaks to draw paintings and records the magnificent mountains, anime style--ar 1:1 --niji 5

a handsome art designer+free

提示词：a handsome art designer, free, male, in an art gallery, the artist sits in front of his own artwork, anime style--ar 1:1 --niji 5

a handsome art designer+creative

提示词：male, a handsome art designer, creative, the sculptor sits in an indoor studio filled with artistic atmosphere, his hand lightly grasping the carving knife, meticulously carving, anime style--ar 1:1 --niji 5

a handsome art designer+free

提示词：a handsome art designer, free, male, takes photos of the scenery, anime style--ar 1:1 --niji 5

a handsome art designer+confident

提示词：male, a handsome art designer, confident, on stage, musicians play music and immerse themselves in it, anime style--ar 1:1 --niji 5

a handsome art designer+elegant

提示词：male, a handsome art designer, elegant,the dancer is dancing,conveying emotions and stories through physical movements and performances, anime style--ar 1:1 --niji 5

a handsome art designer+creative

提示词：a handsome art designer,creative,male,the author sits at the desk in the study, holding a pen and contemplating the plot of the next novel, the room is filled with a serene atmosphere of creation, anime style--ar 1:1 --niji 5

a handsome art designer+fashionable

提示词：male, a handsome art designer, fashionable, actors are sitting in the dressing room of the theater, putting on makeup at the dressing table, anime style--ar 1:1 --niji 5

a handsome art designer+exquisite

提示词：male, a handsome art designer, exquisite, sitting in front of the computer, busy editing graphic design, anime style--ar 1:1 --niji 5

a handsome art designer+exquisite

提示词：male, a handsome art designer, exquisite, sits in the textile room, carefully weaving colorful textiles, anime style--ar 1:1 --niji 5

a handsome art designer+elegant

提示词：a handsome art designer, elegant, male, in the music room, the artist is playing the cello, his expression full of passion and emotion, anime style--ar 1:1 --niji 5

a handsome art designer+elegant

提示词：male, a handsome art designer, elegant, in the dark photography room, the artist is handling work, his camera and darkroom equipment fill the entire room, anime style--ar 1:1 --niji 5

a handsome art designer+elegant

提示词：male, a handsome art designer, elegant, sitting in a café, reading, anime style--ar 1:1 --niji 5

a handsome art designer+exquisite

提示词：male, a handsome art designer, exquisite, the artist sits in a greenhouse, capturing the vitality and vibrant colors of flowers through his paintings, anime style--ar 1:1 --niji 5

a handsome art designer+fashionable

提示词：a handsome art designer, fashionable, male, the artist sits in the corner of the study, with various art albums and magazines on the bookshelf, preparing for the idea of the next masterpiece, anime style--ar 1:1 --niji 5

a handsome art designer+creative

提示词：male, a handsome art designer, creative, on a rainy day, contemplating his creative inspiration, anime style--ar 1:1 --niji 5

a handsome art designer+free

提示词：a handsome art designer, free,male, in a city park, surrounded by open air or cheerful bird songs, anime style--ar 1:1 --niji 5

a handsome art designer+exquisite

提示词：male, a handsome art designer, exquisite, creating exquisite pottery surrounded by displays of flowers and plants, ceramic works, anime style--ar 1:1 --niji 5

a handsome art designer+exquisite

提示词：male, a handsome art designer, exquisite, fashion designer is busy backstage at the fashion show, he checks the model's clothing under the light to ensure that everything is flawless, anime style--ar 1:1 --niji 5

a handsome art designer+creative

提示词：male, a handsome art designer, creative,flipping through the script, thinking about the characters' lines and plot, anime style--ar 1:1 --niji 5

a handsome art designer+creative

提示词：a handsome art designer, creative, male, the director stands at the filming site, directing the actors and production team, anime style--ar 1:1 --niji 5

a handsome art designer+creative

提示词：male, a handsome art designer, creative,using spray paint and paint to create murals, bringing an artistic atmosphere to the city, anime style--ar 1:1 --niji 5

a handsome art designer+elegant

提示词：a handsome art designer, elegant,male, standing on busy street, playing guitar, attracting the attention of passersby, anime style--ar 1:1 --niji 5

a handsome art designer+free

提示词：male, a handsome art designer, free, on the beach, waiting to take the perfect sunset photo and capture the beautiful moments of nature, anime style--ar 1:1 --niji 5

a handsome art designer+creative

提示词：male, a handsome art designer, creative,magician performing wonderful magic on the square, attracting the attention of the crowd, anime style--ar 1:1 --niji 5

a handsome art designer+creative

提示词：male, a handsome art designer, creative, a glass craftsman creates glass artworks in the studio, anime style--ar 1:1 --niji 5

a handsome art designer+confident

提示词：a handsome art designer, confident,male, at an outdoor concert at night, the singer stood on stage with a melodious voice and scattered lights illuminating him, anime style--ar 1:1 --niji 5

a handsome art designer+elegant

提示词：male, a handsome art designer, elegant, floral artist designs exquisite flower beds in an outdoor garden, the colors and aroma of flowers filling the air, anime style--ar 1:1 --niji 5